laserpronet.com
Empowering the Laser Workforce

What we do at LaserPronet

- Screening Exams
- Professional Development Courses
- Professional Growth Plans
- Certifications

Table of Contents

1. Laser Technician Series

Laser Technician I Job Description: Level I Technicians perform final tests on lasers and laser systems to ensure that they fully comply with customer specifications. All tests are completely documented for both internal and external use.

Environment: Test Laboratory.

Requirements: Must understand laser and optics principles and, laser beam performance specifications. Experience testing/evaluating laser beams and working with photonics test equipment is a must. Also, ability to follow prescribed written procedures, directions and strict adherence to best laser lab and manufacturing practices are required.

Minimum Educational Requirements: Candidate must hold a certificate/degree in Laser/Electro-Optics Technology or related discipline

Level 1 Volumes
Volume 1: Basic Laser Optics
Volume 2: Continuous Wave (CW) Laser Principles
Volume 3: Laser Beam Q-switching and Harmonics Generation
Volume 4: Laser Transactional and Performance Specifications
Volume 5: Good Laser Lab and Manufacturing Practices (GLLMP)

Laser Technician II Job Description: Level II Technicians assemble and troubleshoot common optical problems in laser systems. They perform final tests on the systems to ensure that they fully comply with customer specifications and all tests are completely documented for both internal and external use.

Environment: Manufacturing and Test Lab

Requirements: Must understand Wave and Geometrical Optics, Gaussian Beam Propagation, Nonlinear Optics and how acousto- and electro-optics modulators, optical components and accessories work.

Experience aligning laser systems, troubleshooting laser beam aberrations and working with photonics test equipment is a must. Also, ability to follow prescribed written procedures, directions and strict adherence to best laser lab and manufacturing practices are required.

Minimum Educational Requirements: Candidate must hold a certificate/degree in Laser/Electro-Optics Technology or related discipline.

Laser Technician III Job Description: Level III Technicians assemble, align, burn-in, test, and tune/troubleshoot laser heads until they meet all performance specifications.

Environment: Manufacturing

Requirements: Must understand the fundamentals of solid state laser technology, accessories and support systems. Experience aligning laser systems and cavities/resonators and using photonics test equipment is a must. Also, ability to follow prescribed written procedures, directions and strict adherence to best laser lab and manufacturing practices are required. .

Minimum Educational Requirements: Candidate must hold a certificate/degree in Laser/Electro-Optics Technology or related discipline

Technician IV Job Description: Level IV Technicians support Research and Development (R&D) and, customers. Customer Support/Technical Service Technicians work on deployed lasers and laser systems in support of existing customers. R & D/Engineering Technicians support scientists and engineers improve existing, and also create the next generation laser technologies. Level IV Technicians work under limited supervision.

Environment: Research/Engineering Lab and Field

Requirements: Must have a thorough understanding of solid state laser technologies and support systems, experimentation and research protocols. Experience collecting and analyzing data, troubleshooting/problem-solving, using MS Office to generate test reports and writing technical reports is required. Customer Support/Service technicians may have to travel to customer sites. **Minimum Educational Requirements:** Candidate must hold a certificate/degree in Laser/Electro-Optics Technology or related discipline.

2. Laser Technician Level 1 Volume 5
GLLMP Laser Safety *Self-Test*

A. The Human Eye Response to Electromagnetic Radiation [1-8]
B. Laser Beam Hazards [9-17]
C. Laser Beam Output and Attenuation [17-26]
D. Good Laser Safety Practices [27-50]
E. Non-beam Laser Hazards [51-61]
F. Laser Safety Administrative Controls [62-83
G. Laser Safety Engineering Controls [84-87]

A. The Human Eye Response to Electromagnetic Radiation [1-8]

1. The structure of the eye that acts as an image screen is the_____
   a. lens
   b. cornea
   c. retina
   d. pupil
   e. None of the above

2. The focal point of the lens of the eye is at the _____.
   a. retinal
   b. cornea
   c. pupil
   d. a or b
   e. None of the above

3. Collimated electromagnetic radiation that passes through a positive lens is focused at the _____ of the lens.
    a. focal point
    b. center of curvature
    c. vertex
    d. any of the above
    e. None of the above

4. When a laser beam is focused to a point its _____ increases.
    a. irradiance
    b. fluence
    c. a and b
    d. None of the above

5. _____ electromagnetic radiation reaches the retina of the eye
    a. Near-IR
    b. Visible
    c. X-Ray
    d. a and b
    e. None of the above

6. The human eye can detect electromagnetic radiation from about
    a. 400 to 700 nm
    b. 700 to 1400 nm
    c. a and b
    d. None of the above

7. The human eye can underestimate the intensity of the electromagnetic radiation in the about _____ range.
    a. 400 to 700 nm
    b. 700 to 1400 nm
    c. a and b
    d. None of the above

8. Invisible laser beams can be viewed using _____.
    a. cameras
    b. sensor cards
    c. a and b
    d. None of the above

9. Which part of the body is most vulnerable to permanent laser damage?
    a. Skin
    b. Eye
    c. a and b
    d. Hair
    e. None of the above

10. When laser beam energy is absorbed by human skin tissue it generates _____.
    a. heat.
    b. electricity
    c. a and b
    d. None of the above

11. A laser beam in the near infrared part of electromagnetic spectrum can reach the eye _____.
    a. cornea
    b. retina
    c. pupil
    d. b and c
    e. None of the above

12. After a collimated visible laser beam passes through the eye lens _____.
    a. it is focused to the retina
    b. its irradiance/fluence increases
    c. a and b
    d. None of the above

13. _____ laser beams will be focused to the retina of the eye
    a. Near UV
    b. Near-IR
    c. Visible
    d. b and c
    e. None of the above

14. _____ electromagnetic radiation cannot be detected without the aid of special viewers.
    a. IR
    b. UV
    c. a and b
    d. None of the above

15. A burn to the _____ does repair.
    a. cornea of the eye
    b. skin
    c. retina of the eye
    d. a and b
    e. None of the above

16. Laser eye damage can result in _____ blindness
    a. permanent
    b. temporary
    c. no

C. Laser Beam Output and Attenuation [17-26]

17. No labels are required on _____ lasers.
    a. Class 1
    b. Class 2
    c. Class 3b and higher
    d. None of the above

18. The "Caution" warning sign should be used on _____ lasers.
    a. Class 1
    b. Class 2
    c. Class 3b and higher
    d. None of the above

19. _____ lasers are all visible.
    a. Class 2
    b. Class 3a
    c. Class 3b
    d. Class 4
    e. None of the above

20. The "Danger" warning sign should be used on _____ lasers.
    a. Class 1
    b. Class 2
    c. Class 3b and higher
    d. None of the above

21. The higher the _____ of laser eyewear the higher the light intensity that reaches the eye.
    a. optical density
    b. transmission
    c. a and b
    d. None of the above

22. A laser system may have _____ hazard rating as the laser head within it
    a. lower
    b. higher
    c. identical

23. In order to use laser safety eyewear it must
    a. match the color of the laser beam you are to work on.
    b. have both the optical density (OD) and wavelength(s) covered inscribed on it.
    c. state on it the laser they are designated for
    d. b and c
    e. None of the above

24. Assuming your eyewear is designed to protect your eyes for a laser wavelength, the higher the OD of laser eyewear the _____ protection your eyes get.
    a. less
    b. more
    c. Any of the above

25. The Maximum Permissible Exposure (MPE) is the _____.
    a. amount of laser radiation to which a person's eyes may be exposed without the risk of injury.
    b. maximum number of times a person can be exposed to laser radiation before they can succumb to injury.
    c. a and b
    d. None of the above

26. Laser eyewear is designed to permit laser intensity
    _____ the MPE to reach the user's eyes.
    a. less than
    b. equal to
    c. None of the above

D. Good Laser Safety Practices [27-50]

27. The Nominal Hazard Zone is _____
    a. an area within which injury is possible if exposure exceeds the MPE
    b. any space a laser beam can reach
    c. a and b
    d. None of the above

28. It is a good laser safety practice to wear your laser protective eyewear _____.
    a. in the Nominal Hazard Zone/Area
    b. while aligning a laser beam
    c. while operating a laser
    d. all of the above
    e. None of the above

29. It is a good safety practice to place a laser you are working on _____ eye-level
    a. above
    b. below
    c. at
    d. None of the above

30. It is a good laser safety practice to keep an optical bench _____
    a. free of all unnecessary reflective objects
    b. cleaned before use
    c. All of the above
    d. None of the above

31. It is a good laser safety practice not to _____
    a. look directly into laser beam
    b. direct laser beams toward other people.
    c. view a laser beam without the aid of proper optical instruments.
    d. All of the above
    e. None of the above

32. It is a good laser safety practice to verify that the wavelength of the protective eyewear about to be used matches the _____.
    a. color of the laser beams to be operated.
    b. wavelength of the laser to be used/worked on.
    c. a and b
    d. None of the above

33. It is a good laser safety practice to _____.
    a. conceal shiny jewelry from laser beams
    b. contain laser beams to within one's work bench
    c. a and b
    d. None of the above

34. It is a good laser safety practice to _____.
    a. announce to colleagues your intent to turn on a laser before you do
    b. terminate a laser beam when leaving your work area
    c. a and b
    d. None of the above

35. It is a good laser safety practice to terminate a laser beam by turning the_____.
    a. laser off
    b. laser beam shutter on
    c. a or b
    d. None of the above

36. It is a good laser safety practice to _____ laser eyewear before using it
    a. inspect
    b. clean
    c. a and b
    d. None of the above

37. It is a good laser safety practice to read all laser_____ before entering a lab
    a. warning signs
    b. labels
    c. a and b
    d. None of the above

38. It is a good laser safety practice not to _____.
    a. place your hands, or any part of your body, in the path of the laser beam
    b. look directly at laser beam or its reflections
    c. a and b
    d. None of the above

39. It is a good laser safety practice not to _____.
    a. take off laser eyewear from your eyes while the laser is in operation
    b. point a laser beam at anyone
    c. a and b
    d. None of the above

40. It is a good laser safety practice to _____ during laser alignment
    a. lower laser intensity
    b. minimize the number of people around
    c. a and b
    d. None of the above

41. It is a good laser safety practice to ensure that you have _____ protection when working/operating a laser in the UV and shorter wavelengths.
    a. skin
    b. Eye
    c. a and b
    d. None of the above

42. It is a good laser safety practice to figure out the beam _____ of laser before turning it on.
    a. exit aperture
    b. Path
    c. termination point
    d. All of the above
    e. None of the above

43. It is a good laser safety practice to ensure that _____ at the lab door entrance are working before turning laser(s) on.
    a. interlocks
    b. warning lights
    c. a and b
    d. None of the above

44. Laser alignment eyewear is designed to _____ the beam intensity transmitted to the eyes.
    a. block 100%
    b. increase
    c. decrease
    d. All of the above
    e. None of the above

45. Laser alignment eyewear have _____ that differ from regular laser protective eyewear
    a. optical densities (ODs)
    b. color
    c. frames
    d. All of the above
    e. None of the above

46. It is a good laser safety practice to _____ while aligning laser beams
    a. use attenuators
    b. use camera
    c. lower the beam intensity
    d. all of the above
    e. None of the above

47. Laser protective eyewear is designed to attenuate the beam intensity to _____ MPE levels
    a. equal
    b. below
    c. above
    d. None of the above

48. It is not a good laser safety practice to _____ while fully operational.
    a. move a laser
    b. align optics
    c. a and b
    d. None of the above

49. It is a good laser safety practice to point laser beams away from _____.

    a. windows

    b. door and path ways

    c. a and b

    d. None of the above

50. It is a good laser safety practice _____ the path of a laser beam

    a. not to cross

    b. enclose

    c. a and b

    d. None of the above

E. Non-beam Laser Hazards [51-62]

51. Non-beam hazard can include _____ hazards
    a. fire
    b. electrical
    c. a and b
    d. None of the above

52. Exposure of combustible materials to intense laser beams, such as those from class 3b and 4 lasers, could produce _____.
    a. smoking
    b. fire
    c. a and b
    d. all of the above
    e. None of the above

53. Combustible materials include _____.
    a. clothing
    b. wooden items
    c. plastics
    d. all of the above
    e. None of the above

54. _____ resistant materials are particularly suitable to be used for beam termination or as beam dumpers.
    a. Flame-
    b. Water-
    c. a and b
    d. None of the above

55. Non-beam hazard can include _____ hazards
    a. electrical shock
    b. explosions
    c. a and b
    d. None of the above

56. Clutter created by cords and tubes, associated with lasers/laser support systems, in walkways can make personnel
    a. stumble
    b. fall
    c. a and b
    d. None of the above

57. Potentially hazardous gaseous output can result from laser _____ processes
    a. welding
    b. cutting
    c. All of the above
    d. None of the above

58. US Occupational Safety and Health Administration's (OSHA) has _____ on gaseous output of laser target interaction processes.
    a. permissible exposure limits (PEL's)
    b. maximum permissible exposure (MPE)
    c. all of the above
    d. None of the above

59. Enclosing housing is necessary to contain possible explosions associated with laser _____.
    a. high-pressure arc lamps
    b. filament lamps
    c. a and b
    d. None of the above

60. Enclosing housing is necessary to contain possible explosions associated with laser _____ likely to explode.
    a. targets
    b. optics
    c. a and b
    d. None of the above

61. Hazardous UV radiation is likely to be produced by some laser _____.
    a. discharge tubes
    b. pumping lamps
    c. welding plasmas
    d. all of the above
    e. None of the above

F. Laser Safety Administrative Controls [62-83]

62. While ANSI Z136.1 Laser Safety Standards are recommendations they have also been mostly "carbon-copied" by the
      a. EPA
      b. FDA
      c. OSHA
      d. All of the above

63. Almost all _____ programs include the use of the ANSI Z136.1 Laser Safety Standards
      a. environment protection
      b. food and drug safety
      c. laser safety
      d. all of the above

64. You should always ask your Laser Safety Officer, LSO, for information and guidance in connection with laser safety _____.
      a. information
      b. education and training
      c. a and b
      d. None of the above

65. You should always ask your _____ for proper eyewear before working on a laser.
      a. Company President
      b. Technical Manager
      c. Laser Safety Officer
      d. Any of the above
      e. None of the above

66. A laser cannot be operated without following
      a. Standard Operating Procedures (SOPs)
      b. LSO's permission
      c. All of the above
      d. None of the above

67. You are not allowed to operate or work on a laser without proper _____.
    a. training
    b. appropriate protective eyewear
    c. footwear
    d. a and b
    e. None of the above

68. You should wear laser safety eyewear _____ entering a room/lab with an operational laser or turning on a laser.
    a. before
    b. after

69. Before entering a laser facility, you must _____.
    a. observe all posted warning labels
    b. observe entryway controls
    c. put on safety eyewear
    d. All of the above
    e. None of the above

70. If you noticed that a Caution or Danger sign was no longer posted at the usual or appropriate place in your place of employment you should notify _____
    a. the company president
    b. your technical manager
    c. the company Laser Safety Officer
    d. Any of the above
    e. None of the above

71. If you discover an error in the Standard Operating Procedure (SOP) of a laser you would notify
    a. the company president
    b. your technical manager
    c. the company Laser Safety Officer
    d. Any of the above
    e. None of the above

72. If you are not sure which protective eyewear to use while preparing to use, or turn on a laser, you would consult with _____.
    a. the company president
    b. your technical manager
    c. the company Laser Safety Officer
    d. Any of the above
    e. None of the above

73. If the interlocks on a laser/laser system fail to work you would notify _____.
    a. the company president
    b. your technical manager
    c. the company Laser Safety Officer
    d. Any of the above
    e. None of the above

74. If while you are working with a class 4 laser in a lab with door interlocks, and someone opens the door and the laser fails to halt operation you would notify _____ about the incident.
    a. OSHA
    b. your technical manager
    c. the company Laser Safety Officer
    d. Any of the above
    e. None of the above

75. If you suspect that at least one of your eyes is losing its vision whom would you notify first?
    a. an ophthalmologist
    b. your technical manager
    c. the company Laser Safety Officer
    d. Any of the above
    e. None of the above

76. If you are unable to recall laser safety guidelines whom would you contact for laser safety retraining?
    a. the company president
    b. your technical manager
    c. the company Laser Safety Officer
    d. Any of the above
    e. None of the above

77. If your company purchased a new laser and you are not certain if you could operate/use it you would check with _____.
    a. the company president
    b. your technical manager
    c. the company Laser Safety Officer
    d. Any of the above
    e. None of the above

78. You should always consult with your LSO if you need a laser's _____.
    a. Standard Operating Procedure (SOP)
    b. Alignment Procedure
    c. a and b
    d. None of the above

79. It is always a good laser practice to limit the number of _____ when operating or working on a laser.
    a. people in the room with you
    b. people outside the room viewing through a video system
    c. a and b
    d. None of the above

80. Laser eyewear is designed to protect the user from
_____ laser wavelength(s).
 a. a specific
 b. all

81. Skin protection from laser injury is best achieved by
_____.
 a. skin covers such as clothing
 b. beam blockers along the beam path
 c. a and b
 d. None of the above

82. Most laser accidents happen _____.
 a. when eye wear is not worn.
 b. during alignments
 c. a and b
 d. None of the above

83. Laser beam barriers include laser beam blocking
_____.
 a. curtains
 b. widow covers
 c. a and b
 d. None of the above

K. Engineering Safety Controls [84-87]

84. Laser engineering safety controls include
    a. protective housings/chassis
    b. interlocks
    c. activation switches
    d. all of the above
    e. None of the above

85. Laser engineering safety controls include _____.
    a. prevention of unauthorized human access to the laser interior
    b. cessation of laser operation upon removal of chassis
    c. All of the above
    d. None of the above

86. A class 3b, or higher class, laser must have _____.
    a. interlocks
    b. protective housing.
    c. visible and/or audible signals during emission
    d. A key actuated master control system to prevent operation if key is removed
    e. All of the above

87. All lasers must have labels on their housing and visible on the outside that identify_____
    a. the exit aperture of the laser beam
    b. laser class
    c. output power/energy
    d. all of the above
    e. None of the above

3. Blank Answer Sheet

A. The Human Eye Response to Electromagnetic Radiation [1-8]

1. a b c d e
2. a b c d e
3. a b c d e
4. a b c d e
5. a b c d e
6. a b c d e
7. a b c d e
8. a b c d e

B. Laser Beam Hazards [9-17]

1. a b c d e
2. a b c d e
3. a b c d e
4. a b c d e
5. a b c d e
6. a b c d e
7. a b c d e
8. a b c d e
9. a b c d e

C. Laser Beam Output and Attenuation [17-26]

1. a b c d e
2. a b c d e
3. a b c d e
4. a b c d e
5. a b c d e
6. a b c d e
7. a b c d e
8. a b c d e
9. a b c d e
10. a b c d e
11. a b c d e
12. a b c d e
13. a b c d e
14. a b c d e
15. a b c d e

16. a b c d e
17. a b c d e
18. a b c d e
19. a b c d e
20. a b c d e
21. a b c d e
22. a b c d e
23. a b c d e
24. a b c d e
25. a b c d e
26. a b c d e

D. Good Laser Safety Practices [27-50]

27. a b c d e
28. a b c d e
29. a b c d e
30. a b c d e
31. a b c d e
32. a b c d e
33. a b c d e
34. a b c d e
35. a b c d e
36. a b c d e
37. a b c d e
38. a b c d e
39. a b c d e
40. a b c d e
41. a b c d e
42. a b c d e
43. a b c d e
44. a b c d e
45. a b c d e
46. a b c d e
47. a b c d e
48. a b c d e
49. a b c d e
50. a b c d e

E. Non-beam Laser Hazards [51-62]

51. a b c d e
52. a b c d e
53. a b c d e
54. a b c d e
55. a b c d e
56. a b c d e
57. a b c d e
58. a b c d e
59. a b c d e
60. a b c d e
61. a b c d e
62. a b c d e

F. Laser Safety Administrative Controls [62-83]

63. a b c d e
64. a b c d e
65. a b c d e
66. a b c d e
67. a b c d e
68. a b c d e
69. a b c d e
70. a b c d e
71. a b c d e
72. a b c d e
73. a b c d e
74. a b c d e
75. a b c d e
76. a b c d e
77. a b c d e
78. a b c d e
79. a b c d e
80. a b c d e
81. a b c d e
82. a b c d e
83. a b c d e

K. Engineering Safety Controls [84-87]

84. a b c d e
85. a b c d e
86. a b c d e
87. a b c d e

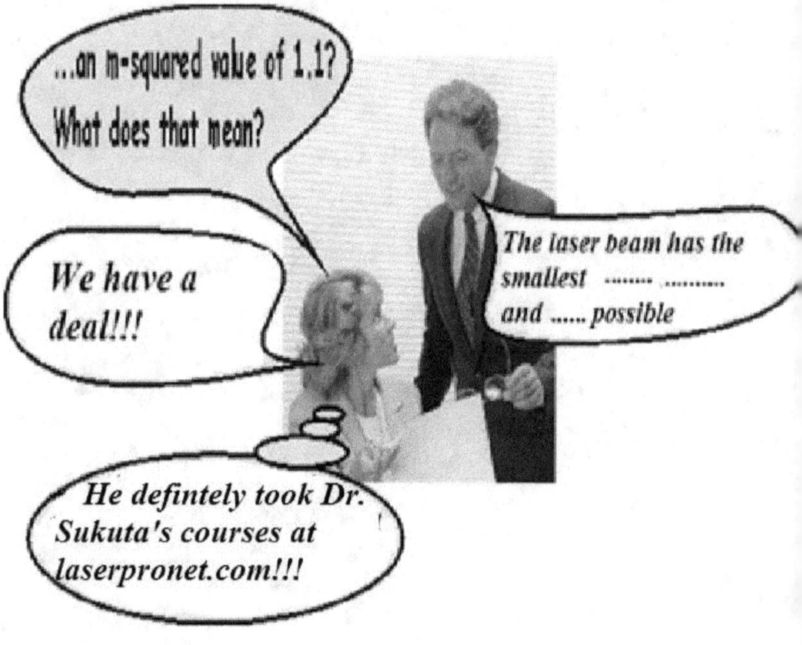

www.ingramcontent.com/pod-product-compliance
Lightning Source LLC
Chambersburg PA
CBHW071832200526
45169CB00018B/1401